Lean Process and Six Sigma Basics

90 Minute Guides

Michelle N. Halsey

Copyright © 2016 Silver City Publications & Training, L.L.C.

Silver City Publications & Training, L.L.C.
P.O. Box 1914
Nampa, ID 83653
https://www.silvercitypublications.com/shop/

ISBN-10: 1-64004-025-0
ISBN-13: 978-1-64004-025-0

Contents

Chapter 1 – Getting Started

During the last couple of decades small, mid-sized and Fortune 500 companies have embraced Six Sigma to generate more profit and greater savings. So what is Six Sigma?

Six Sigma is a data-driven approach for eliminating defects and waste in any business process. You can compare Six Sigma with turning your water faucet and experiencing the flow of clean, clear water. Reliable systems are in place to purify, treat, and pressure the water through the faucet. That is what Six Sigma does to business: it treats the processes in business so that they deliver their intended result.

What is "Sigma"? The word is a statistical term that measures how far a given process deviates from perfection. Sigma is a way to measure quality and performance. The central idea behind Six Sigma is that if you can measure how many "defects" you have in a process, you can systematically figure out how to eliminate them and get as close to "zero defects" as possible. This workshop will give participants an overview of the Six Sigma methodology, and some of the tools required to deploy Six Sigma in their own organizations.

This tutorial is designed to help you in the following ways:

- Develop a 360 degree view of Six Sigma and how it can be implemented in any organization.

- Identify the fundamentals of lean manufacturing, lean enterprise, and lean principles.

- Describe the key dimensions of quality – product features and freedom from deficiencies

- Develop attributes and value according to the Kano Model

- Understand how products and services that have the right features and are free from deficiencies can promote customer satisfaction and attract and retain new customers.

- Describe what is required to regulate a process

- Give examples of how poor quality affects operating expenses in the areas of appraisal, inspection costs, internal failure costs, and external failure costs

- Using basic techniques such as DMAIC and how to identify Six Sigma Projects

- Use specific criteria to evaluate a project

- Discover root causes of a problem

- Design and install new controls to hold the gains and to prevent the problem from returning.

Before you proceed with the tutorial, consider an improvement activity you're about to implement in the work place. The improvement could be about eliminating a certain form, or paper documents, or the use of a certain tool or machine.

Take a moment now to think about your improvement ideas. Keep them in mind as it will help you identify practical applications for the tools and techniques that will be discussed.

Chapter 2 – Understanding Lean

Lean and Six Sigma are buzz-words we hear in business all of the time. Before we get started, let's make sure we all understand just what we mean by "lean" and "Six Sigma".

About Six Sigma

- Six Sigma is a structured, data-driven process of solving critical issues from a business perspective that we haven't been able to solve with current methodology.

- Six Sigma is the single most effective problem-solving methodology for improving business and organizational performance.

- The common measurement scale is called the Sigma capability or Z and is a universal scale. It is a scale like a yardstick measuring inches or a thermometer measuring temperature.

- The scale allows us to compare business processes in terms of the capability to stay within the quality limits established for that process.

- The Sigma scale measures Defects per Million Opportunities (DPMO). Six Sigma equates to 3.4 defects per million opportunities.

What Six Sigma is and is not:

- Six Sigma is not an add-on to normal business activities.

- It is an integrated part of the improvement process.

- Six Sigma is management methodology driven by data.

- Six Sigma focuses on projects that will produce measurable business results.

- Six Sigma is not a standard, a certification or a metric like percentage.

- The central idea behind Six Sigma is that if you can measure how many" defects" you have in a process, you can systematically determine how to eliminate those and approach "zero defects".

Sigma is a value from 1 to 6 that signifies the maximum number of defects per million:
- 1 Sigma = 690,000 defects/million = 31% accurate
- 2 Sigma = 308,537 defects/million = 69.1463% accurate
- 3 Sigma = 66,807 defects/million = 93.3193% accurate
- 4 Sigma = 6,210 defects/million = 99.3790% accurate
- 5 Sigma = 233 defects/million = 99.9767% accurate
- 6 Sigma = 3.4 defects/million = 99.999997% accurate

Six Sigma is about reducing variation.

About Lean

"Lean" means continuously improving towards the ideal and achieving the shortest possible cycle time through the tireless reduction of waste.

- It is focused on eliminating waste in all processes

- It is about expanding capacity by reducing costs and shortening cycle times

- It is about understanding what is important to the customer

- It is not about eliminating people

Examples of Lean Projects:

- Reduced inventory

- Reduced floor space

- Quicker response times and shorter lead times

- Decreased defects, rework, scrap

- Increased overall productivity

History Behind Lean

The phrase "lean manufacturing" was coined in the 1980's and has its roots in the Toyota Production System. (See later in this module)

Most of the basic goals of lean manufacturing are common sense, and some fundamental thoughts have been traced back to the writings of Benjamin Franklin.

Henry Ford cited Franklin as a major influence on his lean business practices, which included Just-in-time manufacturing. The founders of Toyota designed a process with inspiration from Henry Ford and their visits to the United States to observe the assembly line and mass production that had made Ford rich. The process is called the Toyota Production System, and is the fundamental principle of lean manufacturing.

Two books have since shaped the ideologies of Lean: *"The machine that changed the world"* (1990) and *"Lean Thinking"* (1996).

Toyota Production Systems

The Toyota Production System (TPS) is a mindset and management system that embraces continuous improvement. TPS organizes manufacturing and logistics, including interaction with suppliers and customers. Originally called "Just in Time Production," it builds on the approach created by the founders of Toyota. TPS revolves around 5 simple steps:

Step 1: Define Value of your product > Make it according to Customer needs and Customer Defined

Step 2: Identify Value Stream of your product > Follow the product and identify unnecessary actions

Step 3: Study the Flow your product > Eliminate All Waste

Step 4: Make only what the customer orders > Produce Just In Time for Demand

Step 5: Strive for Perfection > Continuous Improvement. Good enough is never enough.

The Toyota Precepts

The five methods defined by Toyota contain some basic principles:

Method 1 - Challenge: Form a long-term vision, meeting challenge with courage and creativity to realize your dreams.

- Create Value through Manufacturing and Delivery of Products and Services
- Nurture a spirit of Challenge
- Always have a Long Range Perspective
- Thorough Consideration in Decision Making

Method 2 - Kaizen: Improve your business operations continuously, always driving for innovation and evolution.

- Have a Kaizen Mind and Innovative Thinking (See later this module)
- Build Lean Systems and Structure
- Promote Organizational Thinking

Method 3 - Genchi Genbutsu (Go and see): Go to the source to find the facts to make correct decisions, build consensus, and achieve goals at our best speed.

- Genchi Genbutsu (Go and See)
- Lead with Consensus Building
- Create Commitment to Achievement

Method 4 - Respect: Respect others, make every effort to understand each other, take responsibility and do your best to build mutual trust.

- Respect for Stakeholders and community
- Develop Mutual Trust and Mutual Responsibility
- Be Sincere, transparent and open in all Communication

Method 5 - Teamwork: Stimulate personal and professional growth, share the opportunities of development, and maximize individual and team performance.

- Have Commitment to Education and Development
- Have Respect for the Individual; Realizing Consolidated Power as a Team

Chapter 3 – Liker's Toyota Way

In this chapter we will look closer at Toyota's philosophies that have become a spiritual pinnacle of modern manufacturing. "The Toyota Way" is a book about the 14 principles that drive Toyota's culture.

The book was written by Dr. Jeffery Liker, a leading author on lean practices and an expert on U.S. and Japanese differences in manufacturing.

Philosophy

Have a Long-Term Philosophy

Principle 1: Base your management decisions on a long-term philosophy, even at the expense of short-term financial goals.

- In Toyota's vision, the purpose is to work, grow, and align the organization toward a common purpose that is bigger than making money.

- The vision instills the importance of generating value for the customer, society, and the economy. The business and its people must accept responsibility for its conduct and continuously improve its skills.

Process

Principle 2: Most Business Processes are 90% Waste and 10% Value-Added Work.

- Create continuous flow and a process flow to bring problems to the surface.

- Work processes are redesigned to eliminate waste (Muda).

- Strive to cut back to zero the amount of time that any project is sitting idle or waiting for someone to work on it.

- The Heart of One-Piece Flow is called Takt Time (Rhythm in German) *"The rate of Customer Demand."*

Principle 3: Use a pull system to avoid overproduction.

- Provide your customers with what they want, when they want it, and in the amount they want.

- Minimize your work in process and warehousing of inventory by stocking small amounts of each product and frequently restocking based on what the customer actually takes away.

- The Toyota Way is not about Managing Inventory, it is about Eliminating It.

Principle 4: Level out the workload (Heijunka).

- Work like the tortoise, not the hare. This helps achieve the goal of minimizing waste (Muda), not overburdening people, or the equipment (Muri), and not creating uneven production levels. (Mura).

- Level out the workload as an alternative to the stop and start approach of working on projects in batches that is typical at most companies.

Principle 5: Build a culture of stopping to fix problems, to get quality right the first time.

- Build into your equipment the capability of detecting problems and stopping itself. Any employee in the Toyota Production System has the authority to stop the process to signal a quality issue.

- It is OK to stop or slow down to get quality right the first time to enhance productivity in the long run.

Principle 6: Standardized tasks and processes are the foundation for continuous improvement and employee empowerment.

- Although Toyota has a bureaucratic system, the way that it is implemented allows for continuous improvement (Kaizen) from the people affected by that system.

Principle 7: Use visual control so no problems are hidden.

- Included in this principle is the 5S Program - steps that are used to make all work spaces efficient and productive, help people share work stations, reduce time looking for needed tools and improve the work environment:

- Sort: Sort out unneeded items
- Straighten: Have a place for everything
- Shine: Keep the area clean
- Standardize: Create rules and standard operating procedures
- Sustain: Maintain the system and continue to improve it

Principle 8: Use only reliable, thoroughly tested technology that serves your people and processes.

- Use technology to support people, not to replace people. Often it is best to work out a process manually before adding technology to support the process.

- New technology is often unreliable and difficult to standardize. A proven process that works generally takes precedence over new and untested technology.

- Conduct actual tests before adopting new technology in business processes, manufacturing systems, or products.

- Reject or modify technologies that conflict with your culture or that might disrupt stability, reliability, and predictability.

People and Partners

Principle 9: Grow leaders who thoroughly understand the work, live the philosophy, and teach it to others.

- The principles have to be engrained; it must be the way one thinks. Employees must be educated and trained: they have to maintain a learning organization.

- Grow leaders and develop role models from within, rather than buying them from outside the organization.

- A good leader must understand the daily work in great detail so he or she can be the best teacher of your company's philosophy.

Principle 10: Develop exceptional people and teams who follow your company's philosophy.

- Success is based on the team, not the individual. Teamwork is something that has to be learned.

Principle 11: Respect your extended network of partners and suppliers by challenging them and helping them improve.

- Toyota treats suppliers much like they treat their employees, challenging them to do better and helping them to achieve it.

- Have respect for your partners and suppliers and treat them as an extension of your business.

- Challenge your outside business partners to grow and develop. It shows that you value them. Set challenging targets and assist your partners in achieving them.

Problem Solving

Principle 12: You need to go and see for yourself to thoroughly understand the situation (*Genchi Genbutsu*).

Principle 13: Make decisions slowly by consensus, thoroughly considering all options; implement decisions rapidly (*Nemawashi*).

The following are decision parameters:

- Find what is really going on; go and see to test

- Determine the root cause

- Consider a broad range of alternatives

- Build consensus on the resolution

- Use efficient communication tools

Do not pick a single direction and go down that one path until you have thoroughly considered alternatives. When you have picked, move quickly and continuously down the path.

Principle 14: Become a learning organization through relentless reflection (*Hansei*) and continuous improvement (*Kaizen*).

The general problem solving technique to determine the root cause of a problem includes:

- Initial problem perception

- Clarify the problem

- Locate the area or point of cause

- Investigate root cause (5 whys)

- Countermeasure

- Evaluate

- Standardize

Once you have established a stable process, use continuous improvement tools to determine the root cause of inefficiencies and apply effective countermeasures.

Chapter 4 – The TPS House

If TPS is a mindset, then what's holding it all together, is the TPS House.

In this module we look at the TPS house, the blueprint for a Lean Enterprise that has become one of the most recognizable symbols of modern manufacturing. The house represents a structural system of how to view our business and organization: The house is strong if the roof, the pillars, and the foundations are strong. A weak link weakens the whole system. It starts with the goals of best quality, lowest cost and shortest lead time – the roof.

There are two main pillars holding the roof up: Just-in-Time (JIT) and Jidoka. JIT and Jidoka mean never letting a defect pass into the next station and freeing people from machines – automation without a human touch. In the center of the system are people.

The Goals of TPS

The Main goals of the Toyota Production System are to eliminate three types of waste:

- Overburden or stress in the system (Muri)

- Inconsistency (Mura)

- Waste (Muda)

The elimination of waste (Muda) is the most common way to look at the effects of TPS. We will look at Waste in greater detail in Module Seven.

There are four rules to TPS:

Rule 1: All work shall be highly specified

Rule 2: Every customer-supplier connection must be direct

Rule 3: The flow of products and services must be simple and direct.

Rule 4: Any improvement must be made according to the scientific method at the lowest possible level in the organization.

The First Pillar: Just In Time (JIT)

JIT is the left pillar and means to make what the customer needs, when it is needed, in the right amount.

Ideally nothing is produced unless a customer is identified and the product is ordered. This helps in reducing inventories, warehousing and other holding costs.

JIT is not about automation. JIT involves controlling the flow of materials and manpower so that adequat resources are on hand when needed.

The Second Pillar: Jidoka (Error-Free Production)

Jidoka is the right pillar of the house.

It means that when an operator detects an error on an assembly line, they will try solving it themselves. If they cannot correct it themselves, they will call their supervisor. If the supervisor cannot complete the job within the given amount of time, the line will be stopped. The error will be fixed and the line will be started.

If you have no solution to the problem, you will not be able to continue with manufacturing. So solving problems becomes a must.

Traditionally, stopping the manufacturing line is treated as a crime, something you should not do at all.

The TPS view is that if you are not shutting down the line, you have no problems. All manufacturing plants have problems. So you must be hiding problems. TPS wants the problems to surface so that the process can be improved. Changing the mentality is the key to implementing Jidoka in an organization.

Kaizen (Continuous Improvement)

Kaizen is a Japanese term that means continuous improvement. With Kaizen, good enough is never enough. No process is ever perfect.

Kaizen aims to eliminate waste in all systems of an organization through improving standardized activities and processes.

The continuous cycle of Kaizen activity has seven phases:

Phase 1: Identify an opportunity

Phase 2: Analyze the process

Phase 3: Develop an optimal solution

Phase 4: Implement the solution

Phase 5: Study the results

Phase 6: Standardize the solution

Phase 7: Plan for the future

The following are some basic tips for doing Kaizen:

- Replace conventional fixed ideas with fresh ones.

- Start by questioning current practices and standards.

- Seek the advice of many associates before starting a Kaizen activity.

- Think of how to do something, not why it cannot be done.

- Don't make excuses. Make execution happen.

- Do not seek perfection. Implement a solution right away, even if it covers only 50 percent of the target.

- Correct something right away if a mistake is made.

The Foundation of the House

The foundation of the TPS house is called Heijunka and means "leveling". Heijunka is a method for reducing waste .The principle is to produce adequate goods at a steady rate, to allow further processing to be carried out at a constant and predictable rate. This stabilization will prevent big spikes in production and hold inventory to a minimum.

Because customer demand fluctuates, two approaches have been developed in lean: demand leveling and production leveling through flexible production.

Chapter 5 – The Five Principles of Lean Business

In this chapter we'll look closer at five great principles that are also known as Womack's Principles. Benchmarking of automotive production facilities, Womack developed a set of five principles that form the basis lean enterprise implementation. The results were published in the book "Lean Thinking", a milestone in Lean Management.

Value

The value of a product or service can only be understood from the customer's point of view. We call this "the voice of the customer."

You need to consider the several voices of the customer:

- What are we falling short of meeting customer needs?

- What are the new needs of customers?

- Voice of Market – Are we ready to adapt to trends?

- Voice of Competitors – Are we behind?

- Voice of Internal Sources – Defects, delays?

- Voice of Employee – Concerns in organization?

Value Stream

- First step in removing non-value added steps from a process is to map the process, following the actual path taken by the part in the plant.

- Walk the full path yourself (Genchi Genbutsu).

- Draw the path on a layout and calculate the time and distances traveled (aka "spaghetti diagram").

Flow

"Flow" means that when your customer places an order, this triggers the process of obtaining raw materials needed just for that customer's order. The raw material then flow immediately to supplier plants, where workers immediately fill the order with components, which flow immediately to a plant, where workers assemble the order, and then the completed order flows immediately to the customer.

Pull

- Push is a traditional manufacturing philosophy - to produce based on estimated forecast of demand.

- The opposite of Pull production is Push production.

- In Pull production, the customer demand instance triggers a part being pulled from upstream.

- Using the Pull philosophy each operation only pulls product from its prior operation when real demand exists at the downstream operation. This results in a continuous flow.

- This will result in many positives for the organization ranging from reduced cycle time, to reductions in inventory to improved customer service levels.

Seek Perfection

Lean perfection is the result of:

- Identifying Value from the customer perspective

- Eliminating waste by mapping the process

- Moving from batch production to Pull Production

- Continuous Improvement within these 3 areas

Chapter 6 – The First Improvement Concept (Value)

The Kano model is a theory of product development and customer satisfaction. When introduced in the 80's the model challenged traditional Customer Satisfaction Models that More is better, i.e. the more you perform on each service attribute the more satisfied the customers will be.

The Kano model assumes that every customer has a unique preference, and that a product has attributes that have different values to different customers.

Basic Characteristics

Kano categorizes the attributes of a product into three classes:

Class 1 - Must-be attributes: Represents basic musts or functions expected of a product or service. When present they are neutral, when absent they dissatisfy consumers.

Class 2 - Performance attributes: Directly linked to voiced demands of customers. Relative to quality and their willingness to pay. Enhances satisfaction.

Class 3 - Surprise and delight factors: These factors excite the customer and satisfy a latent need. Their presence increases satisfaction, their absence does not decrease it. Source of differentiation.

Satisfiers

To identify the "satisfiers" customers are looking for, there is a number of ways to gather data:

- Competitive Analysis

- Interviews, Surveys

- Search Logs

- Usability Testing

- Customer Forums

Delighters

To seek business opportunities and areas of improvement there are a number of resources available including:

- Field Research

- Marketing/Branding Vision

- Industrial Design

- Packaging

- Call Center Data

- Site Logs

From the input gathered, brainstorm a list of features and functionality for the product you intend to improve.

Applying the Kano Model

The Kano classification process is straight forward and simple:

- Identify the Voice of the Customer

Analyze and rank the voice of the customer. Then rank into categories called Critical to Quality Characteristics (CTQs):

- Dissatisfied – Must be – Cost of Entry

- Satisfier – More is better – Competitive

- Delighter – Latent Need – Differentiator

- Evaluate Current Performance

Chapter 7 – The Second Improvement Concept (Waste)

Muda is the waste and work that does not add any value to the product and that the customer would not pay for if given a choice. All waste has a cost that is direct loss to the company. In Lean and Six Sigma manufacturing, the focus is especially on eliminating three types of waste:

- Muda (Waste)

- Muri (Strain/Overburden)

- Mura (Unevenness)

Now let's learn how we can eliminate waste!

Muda

Muda is the waste or work that does not add any value to the product. There are seven kinds of Muda in the Toyota Production System:

- Unnecessary Motions

- Waiting for work and materials

- Transportations

- Overproduction

- Processing

- Inventories

- Corrective Operation (rework and scrap)

Mura

Mura is the variation in the operation of a process not caused by the end customer. It is the Irregular, Inconsistent, Uneven, and Unbalanced work on machines.

A typical example of Mura is when employees are rushing production all morning only to stand around and do nothing later in the day.

Muri

Muri means putting excessive demand on equipment, facilities, and people caused by Mura and Muda. Muri is pushing a machine or person beyond natural limits. Overburdening people results in safety and quality problems. Overburdening equipment causes breakdowns and defects

Examples of Muri include pushing too hard, lifting heavy weight or repeating a tiring action.

The New Wastes

In addition to the three basic types of waste, lean principles have identified other sources of waste in businesses:

- Waste of untapped human potential.

- Waste of inappropriate systems

- Wasted energy and water

- Wasted materials

- Waste of customer time

- Waste of defecting customers

- Waste of unused creativity

Chapter 8 – The Third Improvement Concept (Variation)

If you toss a coin, what's the chance it lands on heads? If you toss a coin ten times, you expect five heads and five tails. If you toss the coin ten times, and do it over and over again, the output varies. The extent to which your experience deviates from expectation is the extent to which variation has occurred.

If you measure something that occurs many times it's going to vary around an average – or mean-value.

Variation is deviation from expectation. The size, trends, nature, causes, effects, and control of this variation is the center of Six Sigma methodology.

Describing variability over a period of time helps us understand how the system is working and to predict how it will continue to work in the future.

Common Cause

When variation is produced by the system itself it is known as common cause variation. You can act to reduce common cause variation but you cannot eliminate it.

- These include human systems, for instance how long it takes to process a credit card application.

- The amount of a time a mechanic takes to change the oil in a car

- An officer writing a speeding ticket

Special Cause

This type of variation is directly caused by something special.

If the mailman comes to your house at 11:30 each day but he gets a flat tire and doesn't come until noon, that's a special cause of variation.

If the network system went down that is running your credit card application, that's a special cause.

Tampering

A common mistake is trying to improve a process by adjusting it when it does not need adjustment. The term for this mistake is called tampering. Tampering usually occurs when the measured outcome does not meet our external performance targets. Examples of external performance targets can be financial targets, and demands to meet monthly quotas.

- Tampering is the adjustment of a stable process.

- If the process is not statistically stable, its performance is unpredictable.

- Tampering leads us to respond to a false alarm, since false alarm is when we think that the process has shifted when it really hasn't.

Structural

These changes are a result of regular, systematic changes in output. They will be more pronounced over the long term and through seasonal patterns. These changes are normal and reflect an ebb and flow of normal production being influenced by the environment.

Chapter 9 – The Fourth Improvement Concept (Complexity)

A product or service that is very complex adds more non-value, higher costs, and more work than processes that are slow or of poor quality. In other words, the complexity of something is more expensive than something that is lower quality or produced in a lower speed.

What is Complexity?

Complexity in Six Sigma means non-value added high cost manufacturing processes. The Law of Complexity and Cost adds more non-value-added cost and work than either poor quality (low Sigma) or slow speed (un-Lean).

What Causes Complexity?

There are two significant contributors to complexity:

Complexity escalates under differentiation, and occurs when we strive to develop a variety of offerings, features, and attributes. As an example, consider a cell-phone manufacturer with a growing number of cell phone models in its portfolio. Each model will require its own R&D, Marketing, and Support.

Complexity escalates under sheer volume of back-operational work. Consider the production of a jet plane which involves hundreds of thousands of engineering specifications and processes that need come together for a final product.

Both scenarios are always at immense strategic risk when faced by a less-complex competitor.

How to Simplify?

Complexity reduction or elimination of non-value added processes is central to Six Sigma and Lean thinking.

There are two approaches to reducing complexity:

Standardization: Standardizing the internal tasks and components of an offering so that a fewer number of them can be assembled into many different products.

A practical example can be found in the automotive industry. Instead of 8 different vehicles built on 8 different platforms, the manufacturer condenses its engineering designs into one platform. Consider how GM now shares one platform across Chevrolet, Cadillac, Buick, and GMC.

Optimization: Eliminating offerings that generate a loss particularly where you are strategically disadvantaged or see a declining market. Almost every organization has products that refuse to generate profit. These should be removed or re-priced to generate adequate return.

The standardization process achieves low cost without the market penalization that an optimization strategy may suffer.

Chapter 10 – The Fifth Improvement Concept (Continuous Improvement)

By now you will know that in the world of lean, good enough is never good enough. In this module, we're breaking down Continuous Improvement into three basic principles:

Principle 1 - **Challenge**: We have to challenge ourselves every day to see if we are achieving our goals.

Principle 2 - **Kaizen**: Good enough is never enough, no process can ever be thought to be perfect, and thus operations must be improved continuously.

Principle 3 - **Genchi Genbutsu**: A term that means "Going to the source" to see the facts for oneself and make the right decisions, create consensus, and make sure goals are attained at the best possible speed.

The PDSA Cycle (Plan, Do, Study, Act)

PDSA is a way to test out improvements on a small scale before implementing them across the board. It will give you the opportunity to see if the proposed change will work. Here's how:

- **Plan the change:** Establish the objectives and processes necessary to deliver results. Set an expected output focus.

- **Do implement the change on a small scale:** Chose a small group of people to test the change.

- **Study the results:** Measure the new processes and compare the results against the expected results.

- **Act on what was learned:** Analyze the differences to determine their cause. Determine where to apply changes that will include improvement.

The DMAIC Method

Define: Identify and state the practical problem

- Who wants the project and why?

- The Scope of the project or improvement
- Key team members and resources for the project
- Critical Milestones and stakeholder review
- Budget Allocation

Measure: Validate the practical problem by collecting data

- Ensure measurement system reliability
- Prepare data collection plan
- How many data points do you need to collect?
- How many days do you need to collect data for?
- What is the sampling strategy? (i.e. where from, from whom)
- Who will collect data and how will data get stored?
- What could the potential drivers of variation be?
- Collect Data

Analyze: Convert the practical problem to a statistical one, define statistical goal and identify potential statistical solutions.

- How well or poorly processes are working compared with Best Possible (Benchmarking) and Competitors
- Don't focus on symptoms, find the root cause

Improve: Confirm and test the Statistical solution

- Present recommendations to process owner
- Pilot run
- Formulate Pilot Run
- Test Improved Process (Run Pilot)
- Analyze Pilot and results
- Develop implementation plan
- Prepare final presentation
- Present final recommendation to management team

Control: Convert the statistical solution to a practical solution

- How will you maintain gains made?
- Change policy and procedures

Chapter 11 – The Improvement Toolkit

So what happens now? We provide some basic methods and organizational advice for the journey ahead. Six Sigma can be a long journey but with a basic understanding of its methods and tools it will improve your work and personal life.

Gemba

Gemba is Japanese for " *actual place*" and in business terms it is where you create value for your customers through daily work, often the factory floor itself.

The idea is that if a problem occurs, the engineers must go to Gemba, the source, or the root, to understand the full impact of the problem.

There are five rules of *Gemba* management:

Rule 1: When a problem arises, go to the *Gemba* first—don't try to make a diagnosis on the phone.

Rule 2: Check the *Genbutsu*—the relevant objects—because "seeing is believing".

Rule 3: Take temporary counter-measures on the spot to resolve the problem.

Rule 4: Then find the root cause of the problem.

Rule 5: Lastly, standardize procedures to avoid a recurrence

Unlike focus groups and surveys, *Gemba* visits are not scripted or bound by what one wants to ask.

Genchi Genbutsu

Genchi Genbutsu means "go and see for yourself".

It refers to the fact that any information about a process will be simplified and abstracted from its context when reported. This attitude of Genchi Genbutsu is to seek "the place where it actually happens".

By observing the actual process or problem at the actual place where it is occurring, the problem solver is able to obtain actual data or facts, which will improve the chances for a better solution.

This is in contrast to the Western thinking in which many managers make decisions from behind a desk, armed only with second hand information from others.

Womack's Principle

The five-step thought processes for guiding the implementation of lean techniques are easy to remember, but not always easy to achieve:

Step 1: Specify value from the standpoint of the end customer by product family.

Step 2: Identify all the steps in the value stream for each product family, eliminating whenever possible those steps that do not create value.

Step 3: Make the value-creating steps occur in tight sequence so the product will flow smoothly toward the customer.

Step 4: As flow is introduced, let customers pull value from the next upstream activity.

Step 5: As value is specified, value streams are identified, wasted steps are removed, and flow and pull are introduced. Begin the process again and continue it until a state of perfection is reached in which perfect value is created with no waste.

Kaizen

The process of becoming a Kaizen learning organization involves criticizing every aspect of what one does. The general problem solving technique to determine the root cause of a problem includes:

- Initial problem perception

- Clarify the problem

- Locate area or point of cause

- Investigate root cause (5 whys)

- Countermeasure

- Evaluate

- Standardize

A Roadmap for Implementation

A Six Sigma initiative begins with a deployment program from the top down. Individuals must go through the required training to become certified belts, and projects have to be identified.

A Six Sigma initiative occurs in five major stages:

Stage 1: Initialize Six Sigma by establishing goals and installing infrastructure

Stage 2: Deploy the initiative by assigning, training, and equipping the staff

Stage 3: Implement projects and improve performance

Stage 4: Expand the scope of the initiative to include additional organizational units

Stage 5: Sustain the initiative, through re-alignment, re-training, and evolution.

When the organization is ready and trained, it's time for a project!

A Six Sigma project should:

- Have financial impact or significant strategic value.

- Produce results that exceed the amount of effort required to obtain the improvement.

- Require analysis to uncover the root cause of the problem.

- Solve a problem that is not easily or quickly solvable using traditional methods.

- Improve performance by greater than 70 percent over existing performance levels.

The focus of a project is to solve a business problem such as:

- The success of the organization

- Costs

- Employee or customer satisfaction

- Process capability

- Output capacity

- Cycle Time

- Revenue Potential

Additional Titles

The 90 Minute Guide series of books covers a variety of general business skills and are intended to be completed in 90 minutes or less. It is an effective way for building your skill set and can be used to acquire professional development units needed by project managers and other industries to maintain their certification. For the availability of titles please see

https://www.silvercitypublications.com/shop/.

No. 1 - Appreciative Inquiry

No. 2 - Assertiveness and Self Control

No. 3 - Attention Management

No. 4 - Body Language Basics

No. 5 - Business Acumen

No. 6 - Business and Etiquette

No. 7 - Change Management

No. 8 - Coaching and Mentoring

No. 9 - Communications Strategies

No. 10 - Conflict Resolution

No. 11 - Creative Problem Solving

No. 12 - Delivering Constructive Criticism

No. 13 - Developing Creativity

No. 14 - Developing Emotional Intelligence

No. 15 - Developing Interpersonal Skills

No. 16 - Developing Social Intelligence

No. 17 - Employee Motivation

No. 18 - Facilitation Skills

No. 19 - Goal Setting and Getting Things Done

No. 20 - Knowledge Management Fundamentals

No. 21 - Leadership and Influence

No. 22 - Lean Process and Six Sigma Basics

No. 23 - Managing Anger

No. 24 - Meeting Management

No. 25 - Negotiation Skills

No. 26 - Networking Inside a Company

No. 27 - Networking Outside a Company

No. 28 - Office Politics for Managers

No. 29 - Organizational Skills

No. 30 - Performance Management

No. 31 - Presentation Skills

No. 32 - Public Speaking

No. 33 - Servant Leadership

No. 34 - Team Building for Management

No. 35 - Team Work and Team Building

No. 36 - Time Management

No. 37 - Top 10 Soft Skills You Need

No. 38 - Virtual Team Building and Management

www.ingramcontent.com/pod-product-compliance
Lightning Source LLC
Chambersburg PA
CBHW060449210326
41520CB00015B/3887